Gold Pan Mining Company

And Shops

Gold Pan Mining Company

And Shops

Breckenridge, Colorado

By Maureen Nicholls

Quandary Press

Maureen Nicholls

Library of Congress No: 94-66385

ISBN: 0-9641029-0-0

Cover design by Tony Wilson, Wilson & Lass, Breckenridge, Colorado, 80424.

Book edited and prepared for publication by Mary Ellen Gilliland, Alpenrose Press, P.O. Box 499, Silverthorne, Colorado 80498.

To mail order, write or call:

Quandary Press
P.O. Box 566
Breckenridge, Colorado 80424
(303) 453-0536

A Gold Pan Panorama

In 1900 the Gold Pan Mining Company transformed the undeveloped southern outskirts of Breckenridge. The 1859 gold boom town became an industrial hub. The town, on the brink of its third mining boom--gold dredging--welcomed its new industrial capability.

The Gold Pan Company's enterprising mining attempt on the Fanny/Maggie Placers failed to produce the expected riches. But the company's machine shops were a remarkable achievement considering the era and location. Its hydroelectric plants modernized local mines.

After the heyday of mining, Breckenridge's economic future dimmed. Towering rock piles surrounded the Gold Pan's deep mining pit and dominated the south end of town, once again its deserted fringes.

In 1961 a fledgling ski resort and second-home development began to change the near ghost-town status of Breckenridge. The giantic growth in the last thirty years, spurred by multi-million dollar resorts, has shrouded the old mining heritage of the Gold Pan properties and created an impact on Breckenridge never before imagined.

Photography

Mr. and Mrs. John H. Marks were an enterprising couple--he a mining engineer for the Gold Pan Mining Company, she the company's photographer. Most of the photographs reproduced in this book were taken by M. M. Marks. This was an unusual feat for a woman considering the equipment used and the early 1900s era.

Mrs. Marks' photos reveal she descended into the rock-strewn pit with the working crew, climbed to the observation platform on the machine shop's roof, scrambled up Ford Hill to set up her tripod and undauntedly photographed icy scenes when the pit was shut down for winter. Most of her photographs have never before been published.

The pit construction photographs were taken by Mrs. Marks with a 4 x 5 view camera. For another series of photographs she used a Cirkut camera that presents elongated panoramic views. (The wide angle lens created some distortion.) An enterprising darkroom printer chopped these 3 1/2 X 11 1/2 inch negatives down to a 10-inch length to fit an 8 X 10 contact frame. However, judging from the original photographs, this hasn't detracted from the full 11 1/2-inch view.

The negatives and their return to Breckenridge unravel an interesting story. Francis Rizzari of Golden, Colorado, a well known Colorado historian and collector of old photographs, came to Breckenridge in the 1950s to ferret out historic Summit County photos. A Breckenridge old timer gave him the name of John Marks, then a consulting engineer in Denver. A friendship developed which led to Francis Rizzari's acquiring the Marks' negatives of the Gold Pan Company's pit, shops and hydroelectric plants. Rizzari stored and preserved these negatives for twenty years and then gave them to another of Colorado's historic photograph collectors, Richard Ronizo, also of Golden.

Five years ago Richard Ronizo intrigued me with prints of the Gold Pan pit and shops. Curious to learn more about an area only two blocks from my home, I was delighted when he gave me the original negatives to print, research and publish.

Printing these ninety year-old negatives was a challenge. The extremely thin nitrocellulose negatives have curled into a straw shape and are very brittle. My ten fingers barely managed to hold the negatives flat enough to clamp down the clips of an antique contact frame. Unfortunately, the wrinkled emulsion, beginning to peel from the plastic support, made them doubly difficult to handle.

Another source for photos was Robert Gore, the grandson and namesake of the superintendent of the Gold Pan machine shops. He inherited a small photograph album which had been presented to his grandfather by Ben Stanley Revett. Revett pioneered gold dredging in the streams around Breckenridge. The album features the interior of the shops and the Reliance dredge, Revett's first successful gold boat, located in French Gulch.

Other photographs in this book were taken by Otto Westerman, Breckenridge's prolific photographer for thirty years. Resident amateur photographers, Minnie Thomas, Erwin Stewart and Maude Brown, also captured views of this industrial south end of Breckenridge.

The acquisitions grew to a large collection of glass plates and cellulose negatives and original photos. Clarity, variety and human interest were paramount in selection. Lying in storage for ninety-odd years, a cache of Breckenridge's past has been uncovered.

Taken from the lofty scaffolding which extended to Main street, Mrs. John Marks' bird's eye view depicts early 1900 Breckenridge. *Courtesy Greg Thompson.*

Acknowledgements

Writing the Gold Pan history depended on the help and information gathered from many sources and people. For so many friends, new and old, I am grateful.

Richard Ronizo's generosity in allowing use of his original negatives and albums enabled me to attempt this project. I am truly indebted to him for his unselfishness and willingness to share.

Bob Gore of Boston was also generous in loaning photos from his family archives and a treasure vault of Gold Pan shops' interior views.

The Breckenridge Writer's Group who meet at Colorado Mountain College on Wednesday nights didn't stop at just encouraging me--they dragged the writing out of me. "Write like you speak" was a comment I heard often. Aptly led by Jane Morton, this group (and especially Carol Oen) helped more than they'll ever know.

Greg Thompson, whose enthusiasm for Gold Pan history kept me going, was a wonderful resource person in railway matters. Mining expertise was gleaned from French Gulch/B. & B. Mines President, John A. Traylor, who read and suggested revisions in copy. For technical information concerning water flows, I asked assistance of Gary Roberts at the Breckenridge Water Department. Charles Wright, a retired Public Service employee and an enthusiastic guide of Georgetown's Energy Museum, shared his expertise and made suggestions about the hydro plants. Summit County's Treasurer Larry Gilliland helped me interpret tax rolls and dig through legal records. Gordon Heuer was my automobile expert.

Thanks to Jim Nicholls, my husband, for this book's map of the Gold Pan Company's pipeline, ditch and south Breckenridge operation.

I am grateful for interviews granted: Frankie Gore Wilson, a delightfully alert ninety plus-year-old lady gave me insight to the personalities of the Gold Pan story; Francis Rizzari recalled information about the Gold Pan's professional photographer; both Wes and Bud Enyeart shared personal memories of the shops, pit and pipeline; Frank Brown was helpful with his recollections. Any factual errors are not theirs but my own.

To my editor Mary Ellen Gilliland of Alpenrose Press, your creative help was invaluable and much appreciated.

M. N.

Turn of the Century Marvel

"Gold Pan, Colorado"--the conductor announced the last stop on the westbound Colorado & Southern Railway entering south Breckenridge from Denver. In 1900 the Gold Pan Mining Company built world-class machine shops at the junction of Ridge Street, Illinois Gulch and the Breckenridge-Alma Roads. Today only a rock constructed dipping tank on south Ridge Street remains.

The Gold Pan Company's mining pit, known now as the Maggie Pond, lies two blocks west of the placer pipe dipping tank and across Highway 9. Named after a placer, the pond today attracts an occasional muskrat paddling near its shore. In summer two pairs of swans swim among paddle boaters. Ice skaters in winter are equally oblivious to the mining operation dug ninety feet below them over ninety years ago. The pond, a prime town asset, evolved from this mining scar.

Except for photographs, little remains of the Gold Pan Company's water diversion project, the largest pipeline built for mining in the state of Colorado. South of Breckenridge rests the concrete base for its first hydroelectric plant. Gold Pan generated electricity lighted the machine shops, the pit operation and the town of Breckenridge.

A boon to Breckenridge's early 1900s economy, the machine shops and electric plants served Breckenridge well into the twentieth century. But in a short five years the focus of the Gold Pan's operation, mining the pit, was abandoned.

The company's history is chronicled here for the first time. Photographs of this turn-of-the-century marvel of the mining world reveal a captivating Breckenridge story.

Contents

Chapter I

The Gold Pan Pit

Blue River Placers:

"Breckenridge, a camp that has turned out more gold with less work than any camp in Colorado," pronounced the January 20, 1900 *Denver Times*. This hype heralded a massive placer mining operation on the Blue River by the Gold Pan Mining Company, incorporated in 1899. The company acquired seventeen hundred acres of placer mining ground along and to bedrock of the Blue south of Breckenridge. Placer mining uses water to wash the surface ground of a mining claim down to bedrock to retrieve the gold. At the depths the grounds were washed however, "surface" was a misnomer. Some placers reached bedrock at ninety feet deep.

The Gold Pan's placer ground consisted of large acreages of mining claims. The land sale had been negotiated by George West of Chicago. C. L. Westerman, a Civil War veteran and long-time Breckenridge resident and mining investor, transacted the real estate consolidation. His compensation, taken in shares of the company, made him a major stockholder.

The company also acquired other placer land in nearby Illinois Gulch. A prior "Gold Pan" mining and milling company had owned the land and the new company adopted the same name. This former operation was east of the 158 foot-long Colorado & Southern Gold Pan trestle spanning Illinois Gulch. The patented (deeded) ground purchased included the south end of Breckenridge, the Park Addition, and the South Side, Maggie and Fanny Placers.

C. L. Westerman, mining promoter and real estate investor, consolidated the placers south of town. This formed the basis of the Gold Pan Company properties. His brother, Otto, was Breckenridge's longtime photographer.

Along the Blue River and south on the Breck-enridge-Alma Road, today Highway 9, placers named

Chicago investors bought stock in the Gold Pan Mining Company. Management raised $1.4 million. *Courtesy Mark and Karen Vendl.*

the Nellie, Crown, Golden Crown, Gold Nuggets, Leap Year, "96", "97" and Dot were consolidated as part of the Gold Pan Company's holdings. Some of them now make up a portion of the town of Blue River.

Over eons, glaciers had eroded gold bearing rock from the mountains and deposited it in streams forming the placers--pockets of gold that migrated from their original source. The heavy gold, worn away from the mother lode, sifted through the gravels to bedrock lying below. The gravels of the upper Blue River, promising rich alluvial gold deposits, had never been tapped.

From the 1870s to the late 1890s placer operations near Breckenridge worked the steeper gulches and streams flowing into the Blue River. Men operating "monitors" or "giants" shot water at great pressure, washing down the sides of these gulches--one form of hydraulic mining. However the Blue's placers themselves were too flat and lacked dumping area for mining companies to wash the gravels with large nozzles of water. Engineers began solving this problem of placering the open terrain with new methods of hydraulically mining the stream beds. Systems of lifting the gravels from bedrock, where the gold lay, were developed and patented.

The Gold Pan Company's management had planned very thoroughly and counted on years of work to placer mine the seven miles of ground south of Breckenridge. The operation was financed mainly by Chicago capitalists at $1,250,000 and promoters estimated potential profits at $60,000,000 to $80,000,000. Sources vary considerably and Fredrick Ransome's *Geology and Ore Deposits of the Breckenridge District* hints of exaggerations. The prospectus glowed. The elaborateness of the operation was hardly believable even for Breckenridge, the largest placer district in the state of Colorado. The December 23, 1900 *Denver Times* glossed: "More gravels (will be) flying than seen in the district for 30 years."

A churn drill, one of three, tested placer ground for gold--one of the Gold Pan Company's first projects.

Prospecting:

The Breckenridge District mining companies were leaders in developing techniques for core drilling placer ground to test for gold. The Gold Pan Company, using this expertise, spent eighteen months prospecting its acreage by boring 120 holes to bedrock using three churn drills, according to Arthur Lakes' January, 1903 article in *Mines and Minerals*. The core samples were cased with six-inch drive pipe and analyzed as to their gold values. The depth of bedrock averaged sixty feet. Due to glacial deposits, the company speculated (mistakenly) gold lay at all levels from grass roots to bedrock. The *Summit County Journal* (*SCJ* or *Journal*) of April 28, 1900 excitedly announced: "Returns from the pannings show all of the town of Breckenridge is built on rich placer ground." This wouldn't be the last time mining companies tempted the townspeople of Breckenridge with reports of riches underfoot!

George H. Evans:

Much of the Gold Pan Company's dynamism came from its manager George H. Evans, a respected and experienced mining engineer. The Gold Pan Company engineered and placed in operation two of his patented hydraulic elevator systems on the Fanny/Maggie Placers. Over 200 of Evans hydraulic plants had already been installed in placer mines worldwide. This state-of-the-art engineering/mining project, known as the Gold Pan pit, was built in 1901 and 1902. A total of five elevators were scheduled to be built, according to the January 5, 1901 *Denver Times*, but photos show only #1 and #2 materialized. George H. Evans' colleagues and

The Gold Pan Company's engineer George H. Evans and his wife, Birdie, set out for a ride in their circa-1903 Stanley Steamer. This was one of the earlier autos in Breckenridge. It operated on coal oil and achieved a speed of 18-20 mph--weather and street conditions permitting. *Courtesy Robert L. Gore Family.*

the townspeople of Breckenridge referred to him as "George H." because a Mark Evans and a George West also served in company management.

British-born George H. Evans brought to Breckenridge impressive credentials as a leading placer mining expert. He also brought to Breckenridge a Stanley Steamer automobile, although it frequently wasn't drivable on rocky, muddy or snowy Breckenridge streets. A pet monkey accompanied the Evans family, highly unusual in frequently frigid Breckenridge. To entertain in his elaborate British style and be close to his company's business, George H. Evans purchased lots on the northwest corner of Adams Avenue and French Street in 1902. In addition to family living quarters for his son, daughter and wife Birdie, he constructed a ballroom and theater to entertain his guests with one of the earliest home movie projection systems in the state. The ballroom/theater portion was built all the way out to French Street and torn down in the late 1950s. The rest of the ninety-year-old house remains.

The Evans' delighted Breckenridge society with new-fangled blinking electric Christmas lights on a revolving tree. The Gold Pan Company's innovative electrician was considered a wizard in a mountain community still burning old fashioned candles!

Hydraulic Elevator:

The Evans elevator employed a complex system to lift rocks, dirt, gravel and anticipated gold from an excavated hole or pit. This technique was expected to be highly successful in the Breckenridge placers. Two

The home George H. Evans built for his family in 1902 occupies the corner of French Street and Adams Avenue. A Gold Pan electric light fixture still graces its high dining room ceiling.

The large theater/ballroom that George H. Evans built on his home extended to the corner of French and Adams Streets. Torn down in the late 1950s, it was the area's first home projection theater.

Construction of the platform for the Evans elevator system built 1901-02 was hampered by wintry conditions. Four men pose for the camera in this icy scene.

pipes for each elevator were sunk vertically from a platform to bedrock. At bedrock men operating "giant" nozzles washed dirt and rocks from the side of the pit into a wooden box, called a bottom sluice. After the debris was collected, the removal process began. Water from the pipeline flowed under great pressure (at 182 pounds per square inch) down one pipe to a nozzle which threw the water, rocks and sands to the mouth of the second pipe. The second pipe hydraulically lifted the under-8-inch rocks and debris to an elevated sluiceway, making the hoist about 100 feet.

The system's complex elevated sluice operation used special gold-collecting riffles which "gathered the precious yellow stuff," elaborated the May 12, 1902 *Denver Times*. The sluiceway also was engineered and patented by George H. Evans. Cocoa matting and canvas were laid below the riffles in order to trap even the smallest particles of gold. A plate of manganese steel protected the sluiceway from the striking rocks.

Mercury was dripped into the riffle system and amalgamated with the gold. After shutting down the operation for "clean-up," workers collected the heavy amalgam and took it to the retort house behind the office building. This mixture of gold and mercury was placed in an iron pot with a pipe vented to a bucket of water. Under heat, the mercury vaporized and then condensed in the water. The molten gold remaining in the iron pot was poured into brick-shaped ingots and sent by Colorado & Southern Railway (C. & S.) to the Denver Mint. Coarse nuggets of any great size found their way to the office--an enticement for potential investors. Silver nuggets were reported to be common.

The Evans elevator system forced water under tremendous pressure down one pipe. A second pipe hydraulically lifted the rocks, dirt, water and anticipated gold to an elevated sluice.

The photo below is dated May, 1902. Workers hit bedrock at 73 feet the following October.

The elevated sluiceway had riffles lining its bottom to trap the gold brought up from the pit.

Employees of the Gold Pan Company used a rocker/cradle and hose as part of the "clean up" operation to recover gold. The process shown here took place behind the Gold Pan office building alongside the retort house. The pit's gold yield was not enough to keep it operating later than 1905.

The attraction of gold lured all the kids in town to the retort building's doors, probably to the consternation of the trusted Gold Pan employees.

The Blue River's course flowed just west of the pit and the Evans hydraulic elevator. To help counteract the seepage problem that required a pump in the pit, a wooden conveyance, called a flume, was built to contain the Blue. The tailings, discharged into the river from the elevator 20 feet above the ground, created a build-up of rocks and gravel along the river's banks and in its bed. From here north, the Blue River flowed through

The Gold Pan pit began as a small excavation. By 1903, it encompassed several acres, its area enlarged by nozzles of water cutting down its sides. *Courtesy Denver Public Library.*

The cableway performed a dual function: It lifted boulders from the pit and transported heavy pipes and equipment into the pit.

Breckenridge as a muddy, churned river carrying silt to ranches "down the Blue."

In an attempt to keep the excavated rocks out of the Blue and the river in its bed, the Gold Pan Company built an elevated wooden scaffold which extended east to Main Street. A rubber conveyor belt moved the rocks from the sluice and dropped them into an ore trolley. The trolley moved up the scaffold and dumped the rocks below this tall platform. These rocks, excavated from the pit, eventually covered most of the scaffolding. A mammoth rock pile, created by this elevator-sluice-conveyor-trolley, overshadowed the south end of Breckenridge.

The Colorado & Southern Railway tracks ran under the conveyor system and then crossed a trestle to the west side of the Blue River. Most of these trestle pilings were inundated by the rocks.

The pipeline provided water at great force to operate the pit's hydraulic nozzles. Called "giants," their water streams washed debris and gold into a bedrock sluice. From the bottom, the elevator pipe brought materials to an upper sluice for gold recovery.

The elevator sluiceway discharged water and rocks into the Blue River.

Men work to keep the Blue River flowing in its bed. They continued building the wooden flume, far right, to divert water gravitating toward the pit. The Colorado & Southern Railway's trestle pilings are beginning to disappear behind rocks.

The Gold Pan Company took extreme measures to keep the elevator rocks from the Blue River and the Colorado & Southern Railway tracks. This elevated scaffold was soon covered with disposed rocks.

Boulder Removal:

Boulders larger than 8 inches were removed from the pit by "stone boats." Men in the pit heaved the boulders into rectangular metal platforms edged on three sides. A cable was attached from the boat to two 110-foot latticed girders on the surface away from the pit. An engineer, using compressed air produced at the Goldpan shops, tightened the cables. The boat lifted the boulders from the pit and automatically dumped them to the west side, creating a rock pile which grew to a monstrous 72 feet. A stone boat had the capacity of making the 1,000 foot trip in 1 1/2 minutes and could handle 3 tons of boulders.

In July, 1903 tracks were laid and two locomotive cranes lumbered down into the pit, expediting the work of the stone boats. As stated in the June 25, 1903 *Mining Reporter*, the pit was "quite ninety feet deep".

Three of seven original pilings for a C. & S. trestle still stand preserved in the Blue River behind the Bell Tower building near the Maggie Pond.

Working in the wet pit couldn't have been too pleasant even on the warmest of Breckenridge summer days. Lack of water pressure shut down the pit for the winter months.

Heaving rocks into the stone boat was dangerous and frequently wet work for the pit crew.

Management watched as the crew picked away at the rocks in the Gold Pan's mining operation.

Two locomotive cranes, launched in summer, 1903, accelerated the work of the stone boat in removing glacial boulders. A water stream washed dirt into the bottom sluice. Spray from its nozzle appears in lower portion of photo.

Boulders from the pit created a 72-foot pile that towered over the pit's west side for nearly sixty years.

Railway cars on the siding at left brought materials to build the Evans elevator. The C. & S. tracks on the right ran under the elevated conveyor, crossed the Blue and continued to the Breckenridge station on Montgomery Street.

Locomotive cranes also worked to crib up the pit walls, an effort to prevent rocks and debris from falling back onto the crew. A falling boulder caused one fatal accident.

Failure of Pit:

The Gold Pan pit ceased functioning by 1905, according to Dr. Ben Parker's *Gold Placers of Colorado*. The pit had taken a full two years to build and had operated only for little more than two years. One hundred men had labored three shifts daily for the Gold Pan Company. Electric arc lights had illuminated the pit for the night crew. One man lost his life in the pit, crushed by a boulder falling back down on the ill-fated miner. Others were maimed. Immeasurable effort and human toll embraced a two-year operation.

The overpowering reason for the pit's failure was simply the lack of enough gold to pay for operating expenses. But technical problems were encountered. Failure was aggravated by the number and size of boulders, some as large as 6 feet. (One the size of a shed was excavated by the Village at Breckenridge, a resort built years later near the banks of the pit.) Other difficulties were the 73-foot depth of bedrock and inadequate prospecting--grounds barren of gold were worked. The samples from the churn drills had incorrectly indicated that gold lay from surface down to bedrock. Another inhibiting factor was that the pit could not be mined during the winter months due to a lack of adequate year-round water supply.

The Gold Pan's initial projections indicated the entire Breckenridge operation would not need overhaul

The Gold Pan pit, abandoned in 1905, soon filled with water.

The large rock pile threatened to cover the company's elevated conveyor. A mountain of small rocks, measuring eight inches or less, accumulated after the pit's hydraulic elevator lifted them up through its heavy pipes. *Courtesy Robert L. Gore Family.*

nor have major shutdowns for the first twelve years. Optimistically, it had expected to placer mine the Blue for the next forty years! These grandiose plans were discarded in five.

Local hopes for the future of the pit prevailed, as was customary in mining communities. *The Western Grocer*, a magazine devoting its October, 1905 issue to Breckenridge (and nothing to grocery business), reported blithely: "The Gold Pan Mining Company which owns one of the largest and best equipped placer plants in the whole universe" had not been operating because the company is "perfecting plans for re-organizing" and would begin operating again next summer, "prosecuted with renewed and increased vigor." The Gold Pan Company's business was directed by Mr. James Hogan, an attorney in Breckenridge and District Attorney for the Fifth Judicial District. George H.

This panoramic view, taken from the company's conveyor belt, faces southwest to reveal a pit that was full, frozen and shut down for winter.

Evans had moved on to other mining pursuits, possibly back to England.

In 1906 the Consolidated Mining and Electric Company became owner of the Gold Pan Company's pit, according to Ben Parker. The amount of gold this company recovered was not enough to meet costs either. Optimism dissolved for even the strongest promoters. The unmined Gold Pan pit filled with water--never to be mined again.

Dredging Interests:

After the Gold Pan Company abandoned its pit, Breckenridge turned its mining attention to the newest darling in mining circles, dredging. Ben Stanley Revett, a successful dredge mining promoter, took over management of the Gold Pan properties in 1907, according to Erl Ellis in *Gold Dredging Boats around Breckenridge, Colorado*. The Gold Pan Company's shops were equipped to repair dredge breakdowns and supply heavy hardware, needed at Revett's dredging operations

The C. & S., steaming toward Illinois Gulch and Boreas Pass, is dwarfed by towering elevator rock piles east of tracks. During pit excavation, the stone boat dumped boulders at right of train.

in the Swan River and French Gulch areas near Breckenridge.

Mr. Revett was a colorful character around Breckenridge and is euphemistically referred to as the "dredge king." His short stature, imposing weight and operatic voice made for wonderful Revett stories. Swan's Nest, the lovely English-style manor he custom built, overlooks his dredge piles on the Swan and remains as a private residence.

In 1913 the Tonopah Placers Company bought out the Gold Pan's properties. The Gold Pan's right hand man, Ben Stanley Revett, stayed on as consulting engineer for the Tonopah's Company's three dredges. The well-known Nevada mining company acquired Revett's Reliance Dredge in French Gulch along with a dredge in the Swan and one on the Blue River. The placers company also obtained extensive holdings on the Swan River and in nearby French Gulch, in addition to the Gold Pan's lands south of Breckenridge.

The Tonopah Company neglected to gain needed rights-of-way and eventually its dredging schemes lacked support from the townspeople and town fathers alike. As Tonopah Dredge No. 1 entered the northern section of Breckenridge on the Blue River, channel widening caused safety problems. One delicate problem involved waste matter draining into the dredge pond and becoming entrapped into cesspools. This was unpleasant particularly for families living in the northern part Breckenridge, used to the swift-flowing river sweeping sewage away from town.

An October 15, 1942 government moratorium on gold mining shut down Breckenridge's last working dredge. Remaining in the "Skelly" dredge pond as a Breckenridge fixture until 1967, the dredge was finally buried in its own tailings by Breckenridge Lands, Inc. operating a D-9 cat. (BLI is now Lincoln West.)

The controversies surrounding the Tonopah Company did not abate. In 1926 this company began selling out to a locally owned, soon-to-be-bankrupt Royal Tiger Mines Company. Through court orders and stock settlements, the Tiger Placers Company and later the Blue River Company emerged. In the proceedings, the courts divided the Royal Tiger holdings in a north-south division through the middle of Breckenridge. The northern portion was under ownership of B. & B. Mines, the initials for Boston and Breckenridge. The south Breckenridge holdings became the property of the Blue River Company.

Operating the last dredge in Breckenridge, the Blue River Company planned to dredge the "sirens," as dredge boat author Ellis called the Fanny and Maggie Placers. Ambitious talk included "stepping" (dredging) east up Illinois Gulch through the South Side Placer. This optimistic enterprise never materialized. It was impractical due to the larger and larger glacial boulders encountered as they dredged south--one of the problems also suffered by the Gold Pan Company. Fortunately for the aesthetics of the local area, the government's moratorium on gold mining during World War II, effective October 15, 1942, thwarted the plan.

The Abandoned Pit:

The deserted mining pit filled with water from the Blue River, fulfilling a railway company's prophetic promotion. Before the Gold Pan Company arrived on the Breckenridge scene, the Denver, Leadville & Gunnison Railway, circa 1894, printed a promotional/informative map of the town. (In 1898 this company became the Colorado & Southern.) Interestingly, a "Breckenridge Lake," using dammed water from the Blue River, was proposed at the location of the present Maggie Pond. It wasn't unusual for railroad companies of that era to promote tourist-oriented attractions, anticipating ticket sales for excursions into the mountains.

During the many years that the flooded pit was abandoned, it was referred to as "bottomless" and became a worry to Breckenridge parents. The *Journal* occasionally reported stories of children--most certainly forewarned of the pit's cold, currents and danger--having mishaps and being rescued from the brink of drowning. In 1917 the *Journal* reported it as a pond of "good size and dangerous depth."

In a 1960s tape of recorded memories, Burl House spoke of his boyhood in Breckenridge from 1906-1910. He frequently fished in the abandoned pit but never caught anything. Even if he had gotten a bite, he said he probably would have let his willow pole drop in the water, afraid of losing his balance and falling in.

Bud Enyeart has been a Breckenridge local since he was five. In 1929 his family moved to a house on south Ridge Street, a short distance from the pit. He'd been sternly warned to stay away from the dangerous abyss, as the banks were steep and sluffed away when anyone walked too close. But he remembers the area as being a favorite playground, particularly a wooden bin, which had been built for a gravel operation of the town. He also recalls looking down from the tracks on the C. & S. trestle over the Blue, scared of the depths and turbulent river below. Three pilings of that trestle remain today. The present diminished flow of the Blue River is due to water diversions south of Breckenridge by the Colorado Springs water board.

Through Breckenridge's quieter years, trees grew up to accompany the rocks surrounding the slumbering pit. Beavers created the main activity. In the 1950s the Blue River Company leased the Fanny and Maggie to a placer mining group with plans of acquiring riches. However, this negotiation resulted more in legal disputes than recovery of gold.

Then in 1959/1960 a newly formed recreation-oriented Summit County Development Corporation (SCDC), a subsidiary of Rounds and Porter Lumber Company of Wichita, Kansas, bought approximately 5,500 acres in and around Breckenridge, paying a paltry $55.00 an acre. Included in this real estate was the south end of Breckenridge, the Fanny, Maggie, and South Side Placers.

SCDC's major development on the Maggie/Fanny ground involved flattening the towering rock piles left by the stone boats and the conveyor belt/trolley system of the Evans elevator. During the summer of 1917 some of these elevator rocks had been crushed and placed on town streets. The *Journal* reported that a "municipal rock crusher made a big dent in the Gold Pan rock piles" and the Wellington Mine had plans to use it on the French Gulch Road. The remaining tailings were detailed on a map drawn of SCDC's acquisitions. Using the heaviest equipment available, a D-9 cat, the large boulders west of the pit and the remaining elevator rocks were flattened. These towering rock piles had overlooked the town of Breckenridge for nearly sixty years. The pit itself remained virtually untouched for another ten years.

The Maggie Pond:

In 1970 Breckenridge welcomed new developments. A group of mainly Chicago investors, (home of the earlier Gold Pan investors) incorporated as The Breckenridge Company. They purchased 114 acres from Breckenridge Lands, Inc., formerly SCDC. Parts of the Maggie and Fanny Placers were included in this sale. The Breckenridge Company developed the Bell Tower building on the eastern banks of the old Gold Pan pit as a base for its Four Seasons condominium projects.

The Breckenridge Company then joined hands with the Breckenridge Ski Area to take a surprising step. In anticipation of ski area expansion on the mountains to

The Bell Tower building was built in 1971 as headquarters for The Breckenridge Company's development at the base of Peak 9. *Courtesy Bill and Becky Conkling.*

the south, they changed the names of Peaks 8, 9 and 10. Renamed as Breckenridge Mountain, Royal Tiger Mountain and Briar Rose Mountain, the new names were printed on published brochures. A good deal of Breckenridge and Forest Service input brought a sheepish retraction. Renaming geographic locations takes a myriad of federal red tape and locals referred to the mountains only as Peaks 8, 9 and 10 anyway. Peak 9 was opened for the 1971-1972 season under the ownership of Aspen Ski Corporation.

The under-construction ski facility base building near the northwest corner of the pit was named the Royal Tiger building and included the Tiger #1 Cafeteria and the Maggie Bar. Soon the entire base building was known as "The Maggie." The old Gold Pan pit became known as "the Maggie Pond"--a summer and winter attraction for locals and tourists alike.

The 1960-1971 developments in the south end of the mining-town-converted-resort-town were only the beginning of change. Before long the south Breckenridge Gold Pan properties would be transformed into a world class resort.

The Maggie Pond and building delighted tourists and locals twenty years ago as much as today. *Courtesy Bill and Becky Conkling.*

Peak 8 still dominates the Maggie skating rink's western view.

Chapter II

Gold Pan Engineering & Mine Supply Company

Need for Placer Equipment:

The "finest machinery plant in the world," praised the March 22, 1902 *Denver Times*. Rave reports commended the Gold Pan Mining Company's first project, the shops in the south end of Breckenridge. Constructed as a specialized pipe manufacturing plant, it was completed in October, 1900. The company needed placer pipes to divert huge volumes of water to its mining operation on the Fanny and Maggie Placers, the Gold Pan pit, known now as the Maggie Pond. The company planned to save substantially by shipping flat boiler plate via Colorado & Southern Railway to Breckenridge.

Gold Pan Company manager George H. Evans

The Goldpan Shops buildings placement on the south Ridge Street property shows in this northeast-facing photo. The pipe/boiler building lay north. Though it appears connected to the center building (machine shops) due to the photographer's angle and lens, they were two separate buildings. The warehouse, right, curved toward Illinois Gulch. At left the south wall of the office building appears.

purchased up-to-date machinery in Pittsburg to manufacture and repair placer pipe and equipment. The shipment to Breckenridge included a flywheel, pipe benders, lathes, a Pelton water wheel, two 180-horsepower boilers and two dynamos, delivered on twenty company-owned railroad cars.

The Oro Grande placer mining operation near Dillon received pipe by rail from the machine shops and sent equipment needing repair back to Breckenridge. Considered a twin company with the same management and principals, the Oro Grande already had installed an Evans elevator, patented by and under the direction of George H. Evans. This complex elevator system hydraulically lifted gravels 100 feet to the sluiceway where the gold was recovered.

Soon after the shops were in operation, management

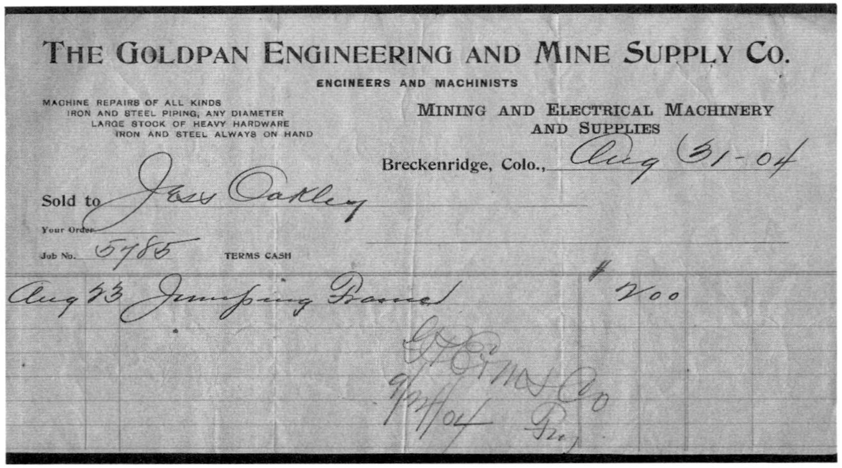
THE GOLDPAN ENGINEERING AND MINE SUPPLY CO.
ENGINEERS AND MACHINISTS
MACHINE REPAIRS OF ALL KINDS
IRON AND STEEL PIPING, ANY DIAMETER
LARGE STOCK OF HEAVY HARDWARE
IRON AND STEEL ALWAYS ON HAND
MINING AND ELECTRICAL MACHINERY AND SUPPLIES
Breckenridge, Colo., Aug 31-04
Sold to Jess Oakley
Your Order
Job No. 5785
TERMS CASH
Aug 23 Jumping Frame $ 2.00

The Gold Pan Mining Company formed the Goldpan Engineering and Supply Company in 1901 as the retail/repair arm of the company's mining business.

The machinery on the left punched holes in steel boiler plate for riveting pipes. The machine shops' original purpose was to manufacture pipe for the Gold Pan Company's placer operations. *Courtesy Robert L. Gore Family.*

realized its highly technical equipment could profitably repair machinery for other mining companies around Summit County. Because its charter didn't allow custom work for the public, the Gold Pan Mine and Placers Company formed the Goldpan Engineering and Mine Supply Company--Goldpan spelled as one word.

In 1901 the Goldpan shops began retailing pipe and heavy hardware along with repairing mine equipment. The shops, located in the Park Addition to the Town of Breckenridge, and the company's south Breckenridge real estate, including the Maggie and Fanny Placers, became property of the new corporation. The sale price of $40,000 was recorded by the Summit County Clerk and Recorder.

Machine Shops Layout:

The Goldpan shops consisted of three large buildings running north and south on Ridge Street and a number of outbuildings. The three-story boiler building/pipe shop was located farthest north. The machine, carpenter, blacksmith shop and powerhouse lay in the middle. A white-fenced walkway with flagpole straddled the roof of this main center building. From this observation deck, Bob Gore, superintendent of the machine shops, set off fireworks for 4th of July celebrations. He also delighted Breckenridge citizens by blasting in the New Year with the shops' steam whistle.

The Goldpan's third building was a 20 by 40-foot two-story warehouse for boiler plate storage. The building was constructed on an angle to the southeast toward Illinois Gulch. The purpose of this odd shape was to accommodate tracks laid to the shops branching off the main C. & S. route. This angle eased loading and unloading materials from the flatbed cars. Although not a house for engines, the curved building resembled and was referred to locally as the "roundhouse."

The $50,000 turn-of-the-century plant included a bath-shower room for the work crew. Shower/changing rooms in mining operations benefited the management

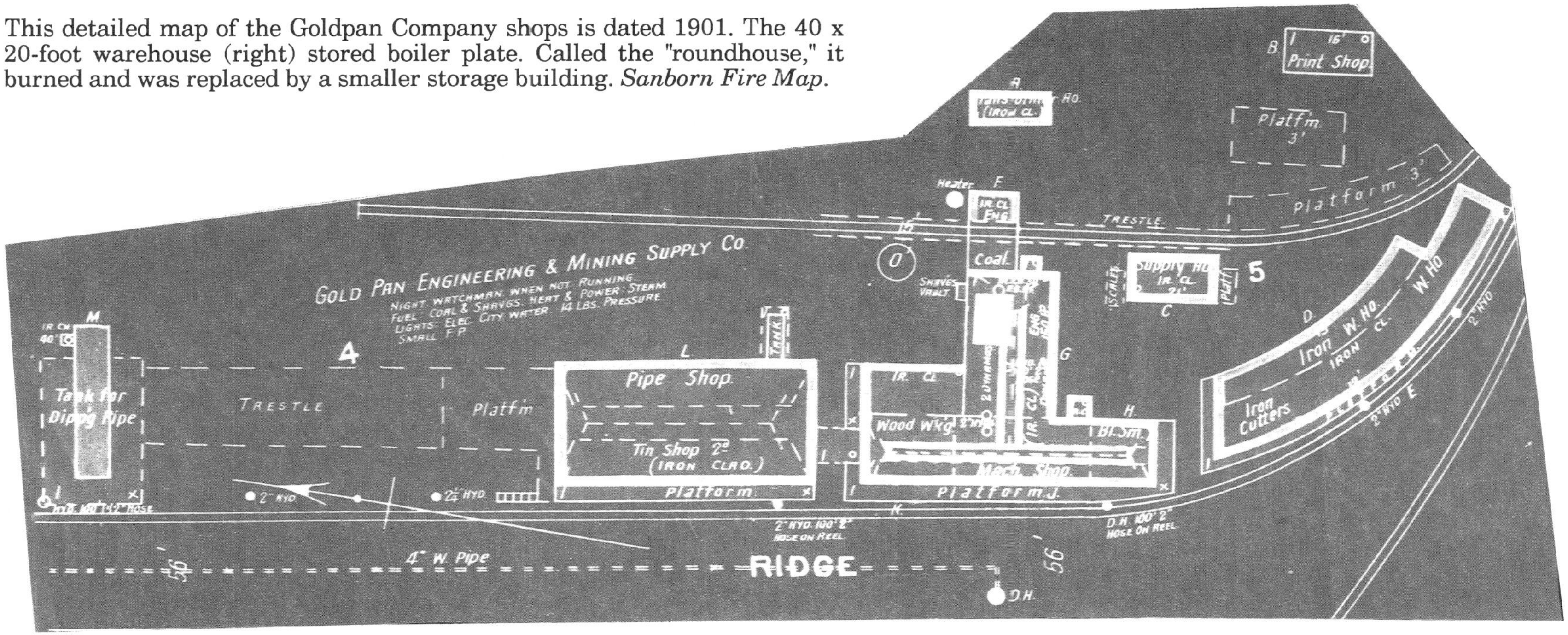

This detailed map of the Goldpan Company shops is dated 1901. The 40 x 20-foot warehouse (right) stored boiler plate. Called the "roundhouse," it burned and was replaced by a smaller storage building. *Sanborn Fire Map.*

The Goldpan's machine shop featured a lath and drill press. Robert Gore, superintendent of the shops, lounges by the potbelly stove. *Courtesy Robert L. Gore Family.*

took turns pushing or riding the cars through the shops and up and down the tracks alongside the buildings. Being friends with the superintendent's children had its advantages.

Above the machine shop was a large room where the Breckenridge "band boys" practiced. Betty Hopkins Royce, whose father was superintendent for the Tonopah Placers Company after the Gold Pan Company sold out, remembers climbing up rickety ladder-type stairs to the attic to play. Scattered around among Gold Pan papers was sheet music, left twenty years before by the band members.

A railroad spur, suitably named "Gold Pan" by the Colorado & Southern Railway, was laid with a double railroad switch to the shops. Trains switched to the rear of the shops ran on an elevated trestle to facilitate dumping coal in large bins. The siding in front of the shops held 31 railway cars and continued north on Ridge Street approximately 1,000 feet (almost to Jefferson Avenue). Flatcars moved pipes to a tank containing asphaltum along this siding.

more than the miners, however. Men required to change clothes before and after mining were not as liable to slip nuggets into their pockets. Looting from mine employers, "high-grading," was an ongoing problem in the mining industry. The twentieth-century shops featured electricity. The interiors were painted with a fire-proof white paint and the exteriors sported the long-lasting iron oxide railroad red. Breckenridge bragged: "The Goldpan shops were the best equipped machine shops between St. Louis and San Francisco! "

Materials and equipment were transported on railway work/hand cars through the three shop buildings. Burl House reminisced on tape about playing with the Gore children on these hand cars. The children

Dipping Tank and Gore Residence:

North of the pipe/boiler building remains part of a dipping tank used by the Goldpan shops. The rock-constructed tank was filled with an asphaltum solution

The Gold Pan Company sponsored a baseball team. Here they join "the boys in the band" to prepare for a July 4 parade and baseball game. The Breckenridge brass band practiced above the machine shop.

which kept the buried placer pipe from rust and corrosion. A coal-fueled vented furnace system, consisting of several brick ovens below, heated the tarry mass. A derrick lifted the 5-foot diameter pipes weighing a hefty 6 to 7 tons and dipped them into the 300-degree asphaltum. The existing rock walls are 7 feet deep by 7 feet wide and 60 feet in length--long enough to hold 48-foot long pipe. A second smaller tank was located behind the boiler/pipe shop.

Wooden platforms storing the uncoated pipe extended from the boiler/pipe shop north along Ridge Street to the tank. After the pipes were dipped, the derrick placed them on additional platforms that reached north and east to French Street. The treated pipe was loaded on flatbed cars from the shops' rear-end tracks and shipped to the mining communities of Summit County. This now rusty pipe, in amazing condition considering its ninety-plus-year history, is scattered in the hills around Breckenridge and still serves a useful life as culverts in Summit County. Easily seen but not so easily moved!

Superintendent Gore and his family lived at 407 South Ridge Street, across the street and a little to the north of the dipping tank. The Gore family spent many happy years in Breckenridge before moving to Leadville in 1918 when Bob Gore became superintendent of the Yak Mine. A clay tennis court, a burro named Gin and her foal Pickles and an overhead trolley with a seat that hung from a cable all entertained the three Gore children: Zoe, Robert, and little Miss Frankie.

A Colorado & Southern Railway car sits on elevated tracks behind the power house at shops rear. The white stairs and walkway leading to the center building's roof is where many of the Gold Pan Company's panoramic photographs were taken.

At Gold Pan, the main line of the C. & S. headed west through the South Side, Maggie and Fanny Placers and over the Blue River. A double switch angled north to the shops. One spur was laid behind and east of the shops. Another spur ran along Ridge Street west of the buildings. Above: Colorado & Southern Railway flatbed cars await departure. They are loaded with 5-foot diameter pipe destined for the Gold Pan pipeline. *Otto Westerman, photographer.*

Massive supplies of placer pipe on storage platforms stretch east to French Street. A derrick lifts a pipe from a hot bath of rust retardant solution that coated each placer pipe.

This view of muddy Ridge Street looks north from the boiler/pipe shop. In the middle of the photo a derrick unloads the pipe from the railway siding. The C. & S. siding reached almost to Jefferson Avenue.

Lumber for the Gold Pan Company's scaffolding arrived by rail at the Gold Pan property on south Ridge. The wooden platform north of the pipe/boiler building extended to the rust retardant tank.

The Gore Family's old home proudly resides at 407 South Ridge Street. Frankie Gore poses in this 1910 vintage photo.

Gold Pan Office:

The Gold Pan Mining Company's office was located on south Ridge Street, west of the machine shops and east of Main Street. From these offices a payroll of $14,000 a month was calculated for the one hundred-plus men employed. The June 1, 1901 *Summit County Journal* elaborated on this two-story "handsomely appointed business home." Five offices, a grand hall and a toilet made up the main floor, while guest bedrooms and bath were located on the second. Visiting company personnel and courted investors made this their home while on inspection trips to Breckenridge.

Alice Swisher, daughter of the *Summit County Journal*'s 1892-1897 editor, celebrated her marriage at the Gold Pan's office building, wearing Mrs. Gore's lovely wedding dress. The night before the wedding the popular Gores hosted the rehearsal dinner. An electrician from the shops fashioned a decorated wire horseshoe, complete with blinking electric lights, over the soon-to-be-wedded couple's chairs. Electricity was the main commodity sold by the Goldpan!

Fire hydrants fed by a four-inch water line were strategically placed along Ridge Street to provide protection from much-feared fires. Despite these precautions, Frankie Gore Wilson remembers a devastating fire at the Goldpan shops. The Gore children sat on the office porch steps watching the "roundhouse" burn in approximately 1908. Their mother, Mabel Gore, had just returned from Denver with new handkerchiefs for her husband. The night of the fire Frankie and her brother and sister left their office porch post to wet these handkerchiefs and hand them out to the men fighting the conflagration. A smaller warehouse was soon built at this location to replace the one destroyed.

The Goldpan shops and office became the stopping-off point for entourages of dignitaries with business in Breckenridge. This location was considerably more impressive for detraining than the Montgomery Street depot on the west side of town. One of the Gold Pan's important visitors was the director of the United States Mint, George Roberts, reported the May, 1902 *Denver Times*. The "eminent mining engineer George H. Evans" squired Roberts who examined the placers south of town. Roberts was duly impressed. The paper touted: "His highest expectations are not only realized but surpassed."

The Gold Pan office building on south Ridge Street, decorated for a patriotic holiday, enjoyed showplace status in Breckenridge. The company's retort building stands behind the office.

The building featured five offices on the main floor with bedrooms and bath upstairs. Rolltop desks were standard office furniture.

Visiting dignitaries--possibly investors--are posed on a railway hand/work car. George H. Evans, far left, is standing and gold dredging pioneer Ben Stanley Revett is seated on the far right.

Visiting aristocracy enthused the mining community of Breckenridge in May, 1904. At the invitation of dredging entrepreneur Ben Stanley Revett, a German prince with the lengthy name of Phillip Ernest Furst zu Hohenlohe Schillingsfurst and his princess sister, Elizabeth Pass Hohenlohe, arrived in a special railway car and enjoyed lunch at the Gore residence on south Ridge.

Five-year-old Bob Gore, namesake of the shops' superintendent, plays with a C. & S. hand brake. The special car transported a visiting prince. Sister Frankie Gore sits on the lowest step while a playmate watches the photographer. *Courtesy Robert L. Gore Family.*

Tonopah Placers Company:

By 1913 the Gold Pan Company's main occupation had been furnishing electricity to the town of Breckenridge and mines in the district. In this year it sold its property to the Tonopah Placers Company, a subsidiary of the Tonopah Mining Company of Tonopah, Nevada. This company bought two churn drills, the Gold Pan ditch and pipeline, the Spruce Creek hydroelectric plant on the Leap Year Placer and the electric transmission line into Breckenridge. The Gold Pan's placers south of Breckenridge were also transferred to this out-of-state company, according to Ellis.

The Tonopah Placers Company bought out the Gold Pan Company in 1913 and used the shops for dredge machinery repair. The abandoned flywheel occupies the railway loading platform.

In addition, the Tonopah Company acquired the holdings of the Colorado Gold Dredging Company on the Swan and the French Gulch properties of Ben Stanley Revett's Reliance Dredge Company. The shops, referred to as the "Tonopah" shops, were mainly needed for repairing the company's three dredges. Tonopah No. 1 worked the Blue north of Breckenridge, No. 2 dredged the Swan and the Reliance Dredge in French Gulch was renamed Tonopah No. 3. Ben Stanley Revett came with the package, continuing his expertise as a consulting engineer, now for the Tonopah Company.

Joe Hopkins, the Tonopah Company's general manager, chose to live near his place of business. He catalog-ordered a pre-cut Sears, Roebuck & Co. home for his lot on the southeast corner of Ridge Street and Jefferson Avenue around 1916. It was conveniently delivered to the Ridge Street lot by the C. & S. and proudly overlooks the Ten Mile Range today. Daughter

Breckenridge had its share of factory pre-cut homes. This Sears model was home for the general manager of the Tonopah Placers Company, Joe Hopkins, and his family. It remains on the southeast corner of Jefferson and Ridge.

Early and late snows were commonly photographed in Breckenridge. This snowy September 12, 1912 photo faces south to take in the north side of the boiler/pipe building and Mrs. Ryan's home. In 1929 the Enyeart Family moved to this house. In the 1970s it was moved to the 100 block of north Harris.

Betty writes of the many happy hours spent roller skating on the concrete basement floor. This was one of the first full basements in a Breckenridge house and she reveals she has a hard time visualizing her roller rink as a time-share condominium today.

Joseph, Betty's younger brother, recalls their hanging play trolley, fashioned by men employed at the shops. Betty remembers: "The take-off stand was heavy with machine grease that used to get on our hands and nails." Mining engineer fathers created innovative play equipment on the sloping lots of south Ridge Street!

When the Tonopah Company took over the shops, much of the sophisticated 1900 equipment was already outmoded. Betty Hopkins Royce said she played in abandoned buildings "full of motors and belts long dead."

In 1929 the Enyeart family made the move to Breckenridge from "down the Blue." The oldest of the five boys herded their five best milk cows with plans to start a family dairy in the south part of town. With the $300 fifteen-year-old Wes saved from a summer job with the Forest Service, his parents purchased a log house located between the Gore's past residence and the Gold Pan's old office building. (Panorama Alpine location today.) As their dad drove the family's Model-T south on Ridge Street, little brother Bud envisioned the shops complex as his new home. When the car rolled to a stop

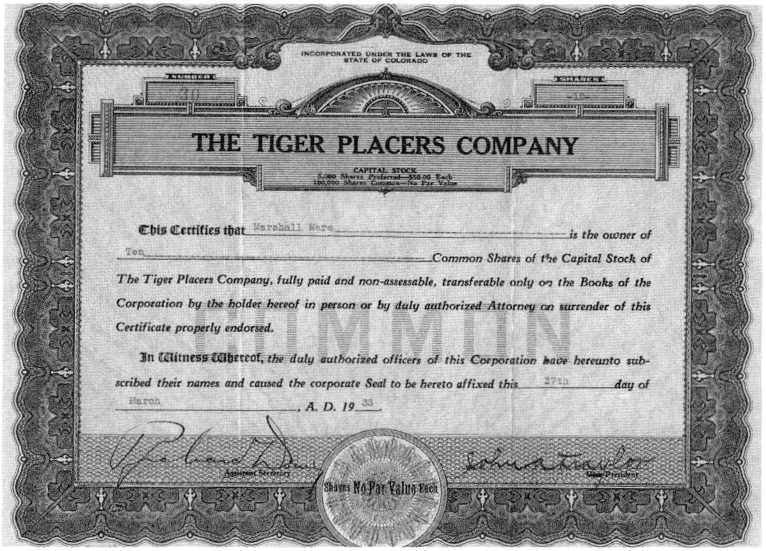

Both the Royal Tiger Mines Company and its reorganized Tiger Placers Company declared bankruptcy. The Blue River Company, formed by disgruntled stockholders, was the last owner of the old Goldpan shops.

at the log house across the street from the pipe storage platforms, he realized his mistake. Bud still recalls the excitement of a five-year-old boy moving from a ranch to across the street from large machine shops. This house was relocated to 105 N. Harris in the early 1970s.

Royal Tiger Mines/Tiger Placers/Blue River Companies:

The Tonopah Placers Company sold out in 1926 to the Royal Tiger Mines Company which also used the repair shops for its newly-acquired dredges. The office building, called "House 5" after the main office of the Tiger, Colorado operation, was sometimes home to the family of Royal Tiger Mines' president, John A. Traylor. The retort house was still used for gold recovery. In fact, son John B. Traylor remembers workers using a lady's silk stocking between the pipe and the surface of the water to keep the water from backing into the iron pot in the retort process.

This company declared bankruptcy two years after purchasing Tonopah's properties. It was reorganized as the Tiger Placers Company. A combination of events--bankruptcy, the Great Depression of the 1930s and the final blow of the government's moratorium on gold mining early in World War II--spelled disaster for Tiger Placers and a third reorganized company.

The Blue River Company, operating the last dredge in Breckenridge until the October, 1942 moratorium, was the final operating owner of the old Goldpan shops. As late as May 1, 1942 the *Journal* praised the machine shops reporting: "Blue River Dredging Company owns machine shops where anything from a darning needle to a threshing machine can be made." This exaggerated picture painted by the *Journal* conflicts with memories shared by both Bud and Wes Enyeart. They remember rundown buildings with much of the machinery already removed and sold to meet financial obligations due to bankruptcy. The Tiger Company had offered Wes a job removing the old machinery in 1933. He turned down that job offer to work on the Continental Dredge just north of Breckenridge. Two job offers in depression-poor Breckenridge were enviable!

In 1940 the Summit County tax rolls listed the value of the real estate, shops, storehouse and office with its furniture and fixtures at $15,060 with taxes levied at $301.20. The obsolete equipment not already sold under bankruptcy court orders or looted and carted away was removed by soldiers on scrap metal drives in W. W. II.

Unpaid taxes in the early 1940s forced Summit County Treasurer Frank Brown to nail tax lien notices to the shop buildings. Brown, Treasurer from 1941-1974, recalled the incident. For a cash-poor Blue River Company, the prospect of lower taxes on undeveloped real estate was incentive enough to schedule the buildings to be demolished. Before Barnett Wrecking/Salvage Company could tear down the weary shops, Summit County's diligent treasurer collected a portion of the past years' delinquent taxes. The old Goldpan shops, only forty years earlier the pride of Breckenridge, came to an ignominious end.

Groundbreaking took place in September, 1961 for the Breckenridge Inn (Breckenridge Mountain Lodge). Construction was just completed when the Peak 8 Ski Area opened on December 15, the date this photo was taken. The lodge floor is lined with historic bricks, removed from two-story vaults in the 1890s county offices in the 100 block of south Main Street.

New Interests:

In 1960 the mostly-vacant property including the shops site, the Maggie and Fanny Placers and much of the south end of Breckenridge, was purchased from the Blue River Company. The new owner was SCDC, known later as Breckenridge Lands, Inc., and presently as Lincoln West, Inc. As support for its Peak 8 Ski Area development (renamed Breckenridge Ski Area), SCDC built a bowling alley, five chalets and the Breckenridge

Inn at the site of the Goldpan-Tonopah-Tiger Placers-Blue River shops. These buildings were ready for Peak 8's opening date of December 15, 1961. Tourist lodging and entertaining visitors in south Breckenridge grew. A forty-room inn's beginnings expanded to a major lodging/resort business on the old mining company's property. The only remnant of the shops is the pipe dipping tank.

How Gold Pan, Colorado has changed!

The rock-constructed placer pipe dipping tank on south Ridge is all that exists of the Gold Pan shops industrial Breckenridge. The remaining asphaltum, warmed to a sticky tar state by the summer sun today, was used to keep pipes from corroding ninety years ago. The tar has been an attractive nuisance to children all these years.

Gold Pan Company manager George H. Evans (left) and shops superintendent Bob Gore (right) pose at a water diversion project headgate.

Chapter III

Gold Pan Pipeline & Goldpan Electric Plants

Demand for Water:

Gold placer mining requires water. A lone miner washed gold with a pan in the stream. Groups of miners ran water through long sluices with riffles to free the gold from dirt. Mining companies employed hydraulic nozzles to wash away the sides of gulches. Floating bedrock dredges ate away the banks of stream beds while digging for gold. A powerful source of water was necessary also to operate the Evans hydraulic elevator at the Gold Pan pit.

Early in the Breckenridge operation, the Gold Pan Mining/Placers Company made plans to divert water to its mining venture on the Fanny and Maggie Placers. The company dug ditches and built a succeeding 8,000 foot pipeline south of Breckenridge. The placer mining invention, the Evans elevator, operated with this diverted pipeline water flowing under pressure down one pipe of the elevator. The water hit the debris at the bottom of the pit, throwing it to a nozzle at the base of a second pipe. This enabled the second pipe to draw the gravels to a sluice above. The pipeline's water pressure also allowed several "giant" nozzles to enlarge the Gold Pan pit by cutting down its banks.

Frederick Ransome's 1911 geological survey of Breckenridge reported that when this water diversion project was completed, the nearly four-mile long ditch and connecting pipeline had a capacity of 6,000 miner's inches with 350 feet of pressure. One miner's inch is an approximate unit of measurement equating the amount of water under pressure flowing through a 1-inch square hole. One miner's inch equals 712.5 gallons per hour. The Gold Pan ditch was 8 feet wide at the bottom, 12 feet in width at the top and 4 feet deep.

Translating this flow into today's terminology, the amount of water the Gold Pan Company had for its peak usage could furnish water for a whopping million people daily in today's standard--an average of 100 gallons per person a day! This flow diminished in the winter--most placer mines had enough water to operate only from April through October. Also, water south of Breckenridge at the turn of the century was not diverted to Colorado Springs as a portion is now.

Location of Pipeline:

The company obtained water rights from the Blue

and its tributaries south of Breckenridge. Seventy-five men using horse-pulled dirt scrapers worked from May to November, 1900. They built a service road to Little Mountain on the South Side Placer, dug trenches, laid pipe and constructed headgates to funnel the stream water.

Seventy-five men built the Gold Pan ditch in summer-fall 1900. Horse-drawn dirt scrapers helped dig the four-foot deep trench. The company obtained water rights from Pennsylvania and Indiana Creeks and the Blue River south of Breckenridge.

The Gold Pan Ditch ran north collecting water from the Blue River and from Pennsylvania and Indiana Creeks to the east. The ditch at this location is now known as the Blue River Trail, popular for cross-country skiing in the winter and horseback riding and biking in the summer. Other sections of the ditch are obliterated by subdivisions and roads or they pass through private property.

Following the contours of the terrain, the ditch flowed into a wooden overflow gate on the Iron Lode east of the Nellie Placer. From there the water was funnelled into an 8,000-foot pipeline, the longest for mining in Colorado in 1900. This pipeline conveyed the water under pressure to the west side of Little Mountain. From that location the pipeline was placed down Illinois Gulch just south of the Colorado & Southern tracks. On the South Side Placer the pipeline split. One section was laid to the shops for a second hydroelectric plant and the other headed to the pit.

A large headgate was constructed below Rocky Point at the beginning of the pipeline. Water pressure from the incline of the five-foot diameter pipe operated the Evans elevator at the Gold Pan pit and generated electricity at the Goldpan shops. *Courtesy Greg Thompson.*

To assure its placer mines enough water to operate, the Gold Pan Company, Arthur Lakes reports, acquired both a 12 and 17-acre site for reservoirs to store water. One of these sites was a small lake at the foot of Rocky

Taken from Boreas Pass near Rocky Point, this photo shows the pipeline in the lower left corner. The pipe continued along the west side of Little Mountain and down Ilinois Gulch south of the C. & S. tracks.

This panorama from Ford Hill looks south into Illinois Gulch, where the rodeo grounds are today. The Gold Pan pipeline appears faintly far left. The Cirkut camera's wide angle caused distortion in the panoramic photos. *Courtesy Greg Thompson.*

A special piece of equipment on tracks lifted and laid the 6 to 7- ton pipe in its seven-foot deep trench. This photo was taken at Gold Pan, near the bottom of Illinois Gulch. Compressed air hoses are visible at lower right.

Point. The other site is unknown and possibly never materialized.

Laying the pipe took a special piece of pipe-laying equipment on tracks. This equipment lifted the 6 to 7-ton pipe and placed it into the prepared canal. Compressed air, sent from the shops, riveted the caulked pipe together. The 5-foot-diameter steel pipeline, coated with rust-retarding solutions at the Goldpan shops, was buried in the 7-foot-deep trench.

Pipeline Removal:

The section of the pipeline laid to the Gold Pan pit was abandoned after 1905 when the pit was shut down. Water still flowed down the pipeline to the shops for generating electricity until the Tonopah Placers Company sold its plants to Public Service in 1925.

Wes Enyeart recalls walking on the abandoned Gold Pan pipeline to the pit to hunt ducks. A successful shoot meant waiting for the currents to float the fowl to shore. His younger brother, Bud, and friends had a different sort of challenge as smaller boys. As a way of proving to be "one of the gang," the boys in Breckenridge dared each other to run alone through the dark pipeline to the end. There was no getting out after a mile and half's scary journey. A grate that kept debris from entering the pipeline forced the youngster to turn around and run back through the dark pipe to the pit.

In 1933 the Tiger Placers Company began selling scrap metal to raise money as required by bankruptcy

Mrs. Marks stood on the observation platform of the Goldpan shops to take this photo of the South Side placer. The Breckenridge-Alma road lies south and the pipeline heads east up Illinois Gulch toward Little Mountain. The Warrior's Mark subdivision, far right, was developed in 1968 with the Blue River flowing through it.

court. Denver salvage companies bid on removal of the pipeline for its scrap metal value. One company, the highest bidder, based its price on the thickness of the pipeline at the pit, a full 7/16 of an inch. Later the company realized that at the beginning of the pipeline, the 5-foot-diameter pipe was only 3/16 inch thick, making a difference in weight and therefore value. The Gold Pan Company's engineers had calculated that water in the pipeline would build up pressure and warranted thicker pipe at the pit than at the headgate. When the salvage company's mistake came to light--to the regret of the Tiger Placers Company and the bankruptcy judge--the pipeline lay dormant for another decade.

In the early 1940s when the United States had entered World War II, a platoon of black soldiers arrived in Summit County to scavenge scrap metal. Uncommon in white Summit County, these soldiers caused quite a stir in the local hangouts when off duty. They salvaged machinery from mumerous mills, mines and inactive dredges, then exhumed and hauled off the buried pipeline. When removed, the pipeline alone provided over 860 tons of steel to help the war effort! Summit County's assessor had valued it at only $500 in the 1942 tax rolls, the last time it was assessed. Blackened timbers at dredge gravesites and at the headgate of the pipeline indicate the acetylene torches used to dismantle equipment ignited accidental fires.

Spruce Creek Plant:

Part of the successful operation of the Gold Pan Company was the production of electricity. The Spruce Creek hydroelectric plant was built in the summer and fall of 1900, according to the November 1, 1900 *Mining Reporter*. The following spring it began to generate electricity for use in the pit and the shops. That summer independent mine companies and the town of Breckenridge began purchasing electricity.

The Spruce Creek hydroelectric plant, 3.6 miles south of the shops, generated electricity transmitted to Breckenridge and to mines south toward Hoosier Pass. Water was brought by ditch to a penstock (pipeline) 500 feet above the plant. This photo faces east.

Located 3.6 miles south of Main and Ridge Street on today's Highway 9, this Spruce Creek plant consisted of the powerhouse, a transformer building and an operator's house. Insulators manufactured by Western Flint Glass Company of Denver (in operation September, 1899-November, 1900--just better than a year) supported the wire that conducted alternating current along the transmission line north to Breckenridge and south to mines toward Hoosier Pass.

Water was diverted from Spruce Creek west of the plant to a ditch and then into a wooden flume leading to a headgate, called a forebay for a hydro plant. From there a buried penstock, a 3-foot pipeline, transported the water to the powerhouse with 500 feet of head. The water, taken by a 30-inch pipe to the water wheel, operated dynamos capable of generating 500 horsepower. After generating electricity, the water was discharged into the Blue River through a concrete tailrace.

The sheet metal shop of the Goldpan Company fabricated this weather vane which was placed on the Spruce Creek hydroelectric plant. *Courtesy Robert L. Gore Family.*

After Spruce Creek's water generated electricity, it emptied into the Blue River through a tailrace. The middle building transformed the direct current into alternating current. Today Highway 9 bisects the cut for the penstock.

A happy foursome celebrates with the Spruce Creek plant and Red Mountain southwest in the background. The Breckenridge-Alma Road, Highway 9, is west of the Spruce Creek hydroelectric plant.

The Breckenridge Electric Company:

Since 1892 the Breckenridge Electric Company, with its coal-fueled steam plant west of the Blue River near Lincoln Avenue, had supplied the town's need for electricity. This company operated from dark to about 2:00 A.M.--a time when most upstanding citizens had little need for lights!

The fickle Breckenridge trustees altered the fate of the Breckenridge Electric Company the night of June 25, 1901. The *Summit County Journal* reported on the town trustees' meeting which ended in travesty. The Goldpan Company presented an electrical bid to the town of Breckenridge for street lights. This bid, twenty dollars a month higher than operator Clem Huntington's Breckenridge Electric Company, stirred more than a little "hot air." A motion by Trustee J. P. Looney to accept the lower bid was defeated by the majority of town trustees in a controversial and highly suspect move. In August, 1901, without a contract, the Goldpan Company began furnishing power to the town of Breckenridge. Eventually the Breckenridge Electric Company went out of business. Individual consumers began purchasing their electricity from the Goldpan Company, which by then furnished competitive rates.

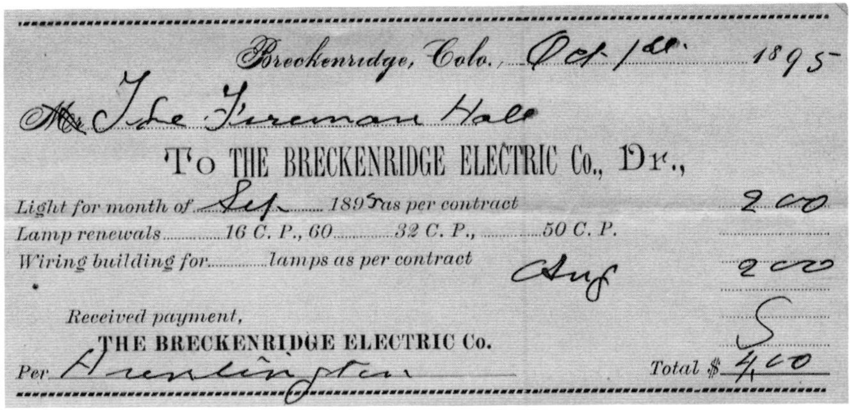

Breckenridge, Colo., Oct 1st 1895

Mr The Fireman Hall

To THE BRECKENRIDGE ELECTRIC Co., Dr.,

Light for month of Sep 1895 as per contract 2 00

Lamp renewals ..... 16 C. P., 60 ..... 32 C. P., ... 50 C. P.

Wiring building for ..... lamps as per contract Aug 2 00

Received payment,

THE BRECKENRIDGE ELECTRIC Co.

Per Huntington Total $ 4 00

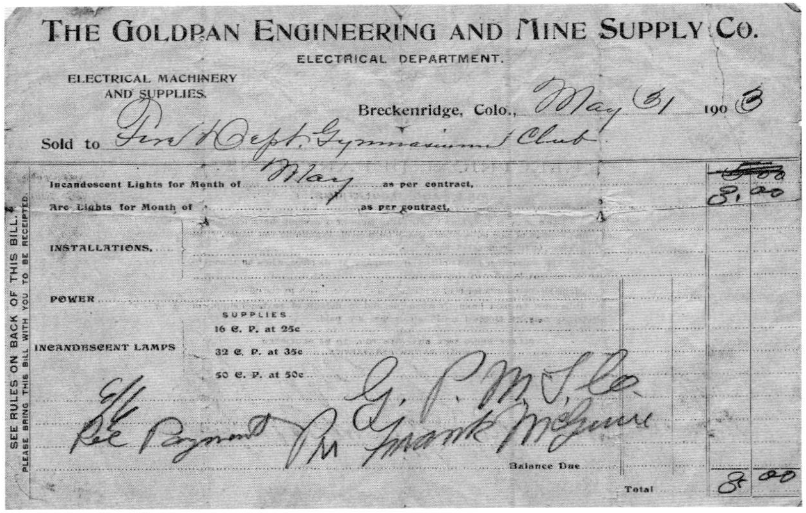

THE GOLDPAN ENGINEERING AND MINE SUPPLY CO.

ELECTRICAL DEPARTMENT.

ELECTRICAL MACHINERY AND SUPPLIES.

Breckenridge, Colo., May 31 1903

Sold to First Dept. Gymnasium Club

Incandescent Lights for Month of May as per contract 8.00

Arc Lights for Month of ..... as per contract

INSTALLATIONS.

POWER

SUPPLIES

INCANDESCENT LAMPS 16 C. P. at 25c, 32 C. P. at 35c, 50 C. P. at 50c

Rec Payment G. P. M. S. Co. Per Frank McGuire

Balance Due

Total 8 00

SEE RULES ON BACK OF THIS BILL.
PLEASE BRING THIS BILL WITH YOU TO BE RECEIPTED.

The Goldpan Engineering and Mine Supply Company began to furnish electricity at the Spruce Creek hydroelectric plant in 1901. Soon its competitive rates put the Breckenridge Electric Company out of business.

Goldpan Shops Electricity:

The Gold Pan ditch and pipeline served a dual purpose: water under pressure operated the Evans elevator and water funnelled to the shops generated electricity at the second Goldpan-owned hydro plant. The Gold Pan Company was faced with a chicken or egg situation when first building in Breckenridge. Water was plentiful for a less expensive hydro power source to run the machinery at the shops. Pipes, manufactured at the shops, were needed to divert the water to the hydro plants to generate the electricity. And so the shops first operated with two coal-fired steam boilers. Spruce Creek furnished electricity in 1901 to the pit, town and mines. By 1903 the hydroelectric plant at the shops was in operation.

Mining companies around Breckenridge began exchanging steam power for cheaper electric power from

Two alternating current generators, driven by belts from the Pelton water wheel, are on the left in the electric room of the Goldpan shops. *Courtesy Robert L. Gore Family.*

the Goldpan Company's two hydro plants. Among those who took advantage of the less expensive power source were the Lightburn Tunnel of the Gold Run Company and the Augusta Mining and Milling Company. They operated electric drills, lights and blowers for ventilation in their tunnels by June and July, 1901 reported the *Summit County Journal*. Power lines were strung to far-flung mines in the Breckenridge District where today roads hardly exist.

Edwin Platt investigated and wrote a report in 1907 on the Wellington Mine, one of Breckenridge's largest producers. He advised management to purchase electricity based on 1 1/2 cents per kilowatt hour from the Goldpan Company. Today electricity is a comparative bargain, considering inflation, at 8 cents a kilowatt hour.

Ellis reports that successful dredging came in conjunction with electrical power. (However, a number of electrocutions occurred on dredges from this little-understood power source.) The Reliance Dredge in French Gulch operated with Goldpan-generated electricity by the summer of 1908. Other dredges were soon converted to electricity for summer operation. However steam power was still the modus operandi in winter, due to unreliable water pressure.

Frankie Gore Wilson recalls a scary incident at the transformer building which was located behind the powerhouse at the shops complex. She and her friend Marvel Sayers were playing paperdolls outside the shops when a summer thunderstorm, complete with lightning, developed. The two little girls scurried into the transformer building to continue their play. She remembers a frightened father, Bob Gore, when he found the children calm, dry and totally unaware of the danger of electrocution they survived.

Hydroelectric plants weren't without operational problems. Clearing out fallen trees and debris from ditches was a regular spring chore. A shortage of water in the stream meant insufficient power generated. Customers were left to limp along anxious for a hard rain or praying for a good snowmelt season.

Along with furnishing electricity for the town of Breckenridge and surrounding mines, the Goldpan shops sold electrical supplies to its customers from a building behind the main shops. Inset: A fire ruined many light fixtures. *Courtesy Greg Thompson.*

Tonopah Company:

In 1913 the Gold Pan Company had sold its shops, electric plants and real estate holdings in Breckenridge and south of town to the Tonopah Placers Company, a subsidiary of a Nevada mining company. Water shortages in 1917 forced the Tonopah Company to purchase power from the Central Colorado Power Company during winter months. This annoyance prompted customers to voice their criticism. The new owners, never considered "local" enough by the Breckenridge populace, had a difficult time being accepted. The *Journal* berated the Tonopah Company for overcharging on electrical rates and suggested that the town should have its own plant. The town's attorney, Carl Kaiser, rendered the opinion that the franchise with the Tonopah Company was illegal as the people had never voted on this supplier. The Tonopah Company countered that the town could indeed buy the power plants.

In 1925 the Tonopah Company began liquidating its Breckenridge assets including the Spruce Creek and Goldpan shops hydroelectric plants. The plants' new owner was the Public Service Company of Colorado, which removed the equipment shortly after purchase. Bud Enyeart remembers only the powerhouse at Spruce Creek standing in the 1930s. Hidden today in the willows, 3.6 miles south on Highway 9, is a little bit of Breckenridge's heritage. Here lie the concrete remains for the foundation and tailrace of the Spruce Creek hydroelectric plant.

This marble control panel came with equipment from the two hydro plants that Public Service Company purchased from the Tonopah Company in 1925. *Courtesy Robert L. Gore Family.*

Legacy

In 1966 when I was a newcomer to Breckenridge (pop. 350), the older residents I met still bragged about the "Tonopah" shops. "The biggest and best equipped west of the Mississippi," they would say with a thumb pointing to the south end of town. But no one explained much about Breckenridge's faded glory.

Alva Springmeyer, a local character honored by a ski run on Peak 8, sold me a copper railway water can with a plaque: "Built at Gold Pan Co. Machine Works, Breckenridge, Colo."

"Used this for a still," he gloated, charging me five dollars. I knew I had a treasured artifact although I didn't know any more about the machine shops. Still intrigued, I wondered what had happened to the machinery and buildings. A long narrow rock vat, spattered with tar, lay a block from my house. What had it been used for? A fixture for so long, the vat never rated a second glance from most residents.

After research into the Gold Pan's history, I'm convinced this 1900s company built the West's most modern machine shops in Breckenridge. Although outdated in a short few years, the shops did justify Breckenridge's sixty-year tradition of bragging.

The old-timers in the 1960s could relate even fewer details about the pit located in the undeveloped southwest part of town. They knew only that it had been "mined."

The Gold Pan operations had fallen quickly from prominence, indicative of the upswings and downturns of a mining community. The impact of the company's placer mining was short and soon forgotten. In fact Frederick Ransome's 1911 geological survey of Breckenridge devotes only a paragraph or two to the Gold Pan Company--just six years after the massive hydraulic mining venture ceased. The bell tolled for George H. Evans' innovative elevator system for lifting bedrock gravels. His last Colorado-built elevator, one of 200 installed worldwide, mined the overrated Maggie and Fanny Placers. No one became rich from the elusive gold including Evans, who disappeared from the Breckenridge scene at the pit's failure.

The miners' dreams of gold faded. Breckenridge survived. Dreams of prosperity became reality in a recreation economy. The mostly delightful climate and desirable location attracted foresighted people. The commitment displayed by George H. Evans and Robert Gore has not been lost. The town's proud heritage has yielded a stronger Breckenridge today.

Bibliography

BOOKS

Digerness, David S. *The Mineral Belt*. Vol. II. Silverton, Colorado: Sundance Publications, Ltd., 1978.

Ellis, Erl H. *The Gold Dredging Boats Around Breckenridge, Colorado*. Boulder: Johnson Publishing Company, 1967.

Fiester, Mark. *Blasted Beloved Breckenridge*. Boulder: Pruett Publishing Company, 1973.

Gilliland, Mary Ellen. *SUMMIT, A Gold Rush History of Summit County, Colorado*. Silverthorne, Colorado: Alpenrose Press, 1980.

Henderson, Charles W. U.S.G.S. Professional Paper #123. *Mining in Colorado*. Washington: U.S. Government Printing Office, 1926.

Parker, Ben H. Jr. *Gold Panning and Placering in Colorado, How and Where*. Information Series #3. Denver: Colorado Geological Survey, Dept. of Natural Resources, 1992.

Poor, M. C. *Denver South Park & Pacific*. Memorial Edition. Denver: Rocky Mountain Railroad Club, World Press, 1976.

Preble, Glen, ed. *Impressed in Time*. Antique Bottle Collectors of Colorado, Inc., Boulder: Johnson Publishing Company, 1987.

Pritchard, Sandra. *Roadside Summit, Part II The Human Landscape*. Dillon, Colorado: Summit Historical Society, 1992.

Ransome, Frederick Leslie. *Geology and Ore Deposits of the Breckenridge District, Colorado*. Professional Paper 75, Department of the Interior, United States Geological Survey. Washington: Government Printing Office, 1911.

INTERVIEWS WITH AUTHOR

Brown, Frank. Breckenridge, Colorado.

Enyeart, Carl Jr. Breckenridge, Colorado.

Enyeart, Wesley and Edna. Canyon City, Colorado.

Traylor, John B. President, B. & B. Mines, Denver, Colorado.

Wilson, Frankie Gore. Denver, Colorado.

Wright, Charles. Institute of Electrical and Electronics Engineers, Inc. Tour Director, Georgetown Energy Museum. Georgetown, Colorado.

PERIODICALS

Lakes, Arthur. "Summit County Placers." *Mines and Minerals*. Vol. 23, No. 6 (Jan. 1903). Denver, Colorado.

Mining Reporter, Vol. 42, (1900); 43- 44 (1901); 45-46 (1902); 47-48 (1903); 49 (1904). Denver, Colorado.

Mining Scientific Press, Vol. 104, No. 26 (June 29, 1912). San Francisco, California.

The Mining World, World Printing Company, 1901. 1850 Lawrence St., Denver, Colorado.

Parker, Ben H. Jr. "Gold Placers of Colorado." *Quarterly of the Colorado School of Mines*, Vol. 69, No 3 (July, 1974) Golden, Colorado.

The Western Grocer, Butcher and Clerk. Denver, Colorado. Vol. 17 (Oct. 1905). W. S. Keller Proprietor, 1853 Lawrence St., Denver, Colorado.

Turnbull, Bell. "Gold Boats of the Swan." *The Colorado Magazine*, State Historical Society of Colorado, Vol. 39, No. 4, (Oct., 1962) Denver, Colorado.

MAPS

George, R. D. Colorado Topographical 1913.

Denver, Leadville and Gunnison Railway Map of Breckenridge, Undated.

Royal Tiger Mines Claims Map, Undated.

T. A. Brown, Placers and Lode Map, 1903.

Ransome, 1911.

Sanborn Map Co., Breckenridge, Summit County, Washington: Library of Congress, Maps and Charts, 1902 and 1914.

NEWSPAPERS

Denver Times: July 20, 1899; Oct. 19, 1899; Jan. 20, 1900; Nov. 12, 1900; Nov. 17, 1900; Dec. 23, 1900; Jan 5, 1901; May 4, 1901; May 22, 1901; May 29, 1901; Sept. 7, 1901; Mar. 22, 1902; Apr. 25, 1902; May, 12, 1902; July 14, 1902; July 21, 1902.

Summit County Journal: July, 1898; Aug. 27, 1898; Oct. 15, 1898; Nov. 26, 1898; Oct. 28, 1899; Nov. 4, 1899; Feb. 3, 1900; Mar. 10, 1900; Apr. 28, 1900; June 2, 1900; Aug. 4, 1900; Nov. 17. 1900; Jan. 20, 1900; Nov. 17, 1900; Jan. 9, 1901; Feb. 9, 1901; Mar. 23, 1901; Apr. 28, 1901; May 4, 1901; June 1, 1901; June 8, 1901; June 22, 1901; June 29, 1901; July 20, 1901; Aug. 17, 1901; Sept. 7, 1901; Nov. 23, 1901; July 8, 1902; Aug. 8, 1903; Apr. 21, 1917; June 7, 1917; Oct, 20, 1917; Dec.

29, 1917; Mar. 11, 1938.

UNPUBLISHED SOURCES

Letters

Hopkins, Joseph E. Jr. Correspondence concerning family property at 400 So. Ridge St., Breckenridge, Colorado, and his memories of Tonopah Placers Company.

Perrin, Mrs. Charles. Correspondence concerning family property at 407 So. Ridge St., Breckenridge, Colorado.

Platt, Edwin H. Report on Wellington Mines, Breckenridge, Colorado. Denver, Colorado. July 28, 1907, University of Missouri Historical Collection, Rolla, Missouri.

Royce, Betty Hopkins. Correspondence concerning family and property at 400 So. Ridge St., Breckenridge, Colorado, and her memories of Tonopah Placers Company.

Documents

Chart of Corporate Development of the Public Service Company of Colorado. Colorado State Planners Commission, April 1, 1938.

Daugherty, John. "Cultural Resources Inventory for Summit County, Colorado." Dillon, Colorado: Summit Historical Society, 1974.

Record Books. Summit County Clerk and Recorder's Office

Tax Assessment Rolls. Summit County Treasurer's Office.

Index